本书由中国科学院数学与系统科学研究院资助出版

数学 24/7

游戏中的数学

〔美〕詹姆斯·菲舍尔　著

李　欣　赵彦超　译

科学出版社

北　京

图字：01-2015-5626号

内 容 简 介

　　游戏中的数学是"数学生活"系列之一，内容涉及掷骰子、大富翁游戏、国际跳棋、纸牌游戏等方面，同时介绍了其中涉及的概率知识等，让青少年在学校学到的数学知识应用到与游戏有关的多个方面，让青少年进一步了解数学在日常生活中是如何运用的。

　　本书适合作为中小学生的课外辅导书，也可作为中小学生的兴趣读物。

图书在版编目（CIP）数据

游戏中的数学/（美）詹姆斯·菲舍尔（James Fischer）著；李欣，赵彦超译.—北京:科学出版社,2018.5
（数学生活）
书名原文：Game Math
ISBN 978-7-03-055693-6

Ⅰ.①游… Ⅱ.①詹… ②李… ③赵… Ⅲ.①数学-青少年读物 Ⅳ.①01-49

中国版本图书馆CIP数据核字（2017）第292431号

责任编辑:胡庆家 / 责任校对:邹慧卿
责任印制:肖 兴 / 封面设计:陈 敬

科学出版社 出版
北京东黄城根北街16号
邮政编码：100717
http://www.sciencep.com

北京汇瑞嘉合文化发展有限公司 印刷
科学出版社发行 各地新华书店经销
*
2018年5月第 一 版　　开本:889×1194 1/16
2018年5月第一次印刷　　印张:4 1/2
字数:70 000

定价：**98.00元**（含2册）
（如有印装质量问题，我社负责调换）

引　言

你会如何定义数学？它也许不是你想象的那样简单。我们都知道数学和数字有关。我们常常认为它是科学，尤其是自然科学、工程和医药学的一部分，甚至是基础部分。谈及数学，大多数人会想到方程和黑板、公式和课本。

但其实数学远不止这些。例如，在公元前5世纪，古希腊雕刻家波留克列特斯曾经用数学雕刻出了"完美"的人体像。又例如，还记得列昂纳多·达·芬奇吗？他曾使用有着赏心悦目的尺寸的几何矩形——他称之为"黄金矩形"，创作出了著名的画作——蒙娜丽莎。

数学和艺术？是的！数学对包括医药和美术在内的诸多学科都至关重要。计数、计算、测量、对图形和物理运动的研究，这些都被融入到音乐与游戏、科学与建筑之中。事实上，作为一种描述我们周围世界的方式，数学形成于日常生活的需要。数学给我们提供了一种去理解真实世界的方法——继而用切实可行的途径来控制世界。

例如，当两个人合作建造一样东西时，他们肯定需要一种语言来讨论将要使用的材料和要建造的对象。但如果他们建造的过程中没有用到一个标尺，也不用任何方式告诉对方尺寸，甚至他们不能互相交流，那他们建造出来的东西会是什么样的呢？

事实上，即便没有察觉到，但我们确实每天都在使用数学。当我们购物、运动、查看时间、外出旅行、出差办事，甚至烹饪时都用到了数学。无论有没有意识到，我们在数不清的日常活动中用着数学。数学几乎每时每刻都在发生。

很多人都觉得自己讨厌数学。在我们的想象中，数学就是枯燥乏味的老教授做着无穷无尽的计算。我们会认为数学和实际生活没有关系；离开了数学课堂，在真实世界里我们再不用考虑与数学有关的事情了。

然而事实却是数学使我们生活各方面变得更好。不懂得基本的数学应用的人会遇到很多问题。例如，美联储发现，那些破产的人的负债是他们所得收入的1.5倍左右——换句话说，假设他们年收入是24000美元，那么平均负债是36000美元。懂得基本的减法，会使他们提前意识到风险从而避免破产。

作为一个成年人，无论你的职业是什么，都会或多或少地依赖于你的数学计算能力。没有数学技巧，你就无法成为科学家、护士、工程师或者计算机专家，就无法得到商学院学位，就无法成为一名服务生、一位建造师或收银员。

体育运动也需要数学。从得分到战术，都需要你理解数学——所以无论你是想在电视上看一场足球比赛，还是想在赛场上成为一流的运动员，数学技巧都会给你带来更好的体验。

还有计算机的使用。从农庄到工厂、从餐馆到理发店，如今所有的商家都至少拥有一台电脑。千兆字节、数据、电子表格、程序设计，这些都要求你对数学有一定的理解能力。当然，电脑会提供很多自动运算的数学函数，但你还得知道如何使用这些函数，你得理解电脑运行结果的含义。

这类数学是一种技能，但我们总是在需要做快速计算时才会意识到自己需要这种技能。于是，有时我们会抓耳挠腮，不知道如何将学校里学的数学应用在实际生活中。这套丛书将助你一马当先，让你提前练习数学在各种生活情境里的运用。这套丛书将会带你入门——但如果想掌握更多，你必须专心上数学课，认真完成作业，除此之外再无捷径。

但是，付出的这些努力会在之后的生活里——几乎每时每刻（24/7）——让你受益匪浅！

目　　录

Contents

1

掷骰子游戏和概率

梅森喜欢玩游戏。无论在哪儿——学校、周末在家、下课后在朋友家、公交车上……只要一有机会，他都想玩游戏。每发现一种新游戏，他都跃跃欲试。

梅森也非常喜欢数学，这是他在学校最喜欢的一门功课。一天，他在课堂上学习了概率，即某个事件发生的可能性。老师说，概率可以用分数表示。

后来，在玩掷骰子游戏时，梅森意识到，掷骰子游戏和概率有很大的关系。他还发现，绝大多数游戏，尤其是需要掷骰子的游戏，至少在一定程度上是以概率为基础的。掷骰子与概率有什么关系呢？下面我们来揭晓答案。

梅森在玩一枚普通的6面骰子，骰子的6个面上分别刻着1到6。他想知道掷出3的概率是多少。

首先，要考虑掷一枚骰子将出现多少种结果（或可能性），这个结果数就是概率分数的分母。

1. 当他掷一枚骰子时，会出现多少种结果呢？

要想得到概率分数的分子，就必须考虑一件事件的发生有几种可能结果。在这种情况下，即考虑梅森掷出3点的可能结果有几种。

2. 掷出3点的可能结果有几种呢？

现在将这两个数字组合成一个概率分数：1/6，于是我们可以说，梅森掷出3点的概率是1/6。

你也能够计算掷骰子时掷出某一类点数的概率。例如，一次掷出偶数的概率是3/6。因为掷一枚骰子有6种结果，掷出偶数有3种可能。

尽可能化简分数，这样才能更深入地理解概率的含义。

3. 3/6的化简结果是什么？你能换一种方式来解释这个概率吗？

梅森的家里还有一些"疯狂骰子"，一枚骰子有12面，他甚至有一枚20面的骰子。用这些骰子，就能够进行一些更为复杂的概率计算。

用一枚12面骰子掷出3点的概率是1/12，用一枚20面骰子掷出3点的概率是1/20。

4. 1/12的概率与1/6的概率相比，是大还是小呢？为什么？与1/20的概率相比呢？

5. 掷一枚20面骰子，掷出的点数正好可以被3整除的概率是多少？你可以先列出这些点数，然后计算其总数。

2

旋转轮盘：更多关于概率的内容

梅森玩得最多的棋盘类游戏是旋转轮盘。玩家转动棋盘上的指针，指针绕圆周转动，停止后指向一个特定数字。

如同掷骰子游戏一样，旋转轮盘也涉及概率。如果轮盘上有6个空格，那么指针指向3的概率就与6面骰子掷出3点的概率是一样的。得到3的次数相同，概率也相同。

总之，只要掌握基本原理，就可以从不同的角度考虑概率的涵义。概率可以表示成分数，也可以表示成小数和百分数。

你已经知道，指针指向3的概率是1/6，下一步要弄明白如何把分数换算成小数。

把分数换算成小数，简单地说，就是用分子除以分母。

1. 概率为1/6，表示成小数是多少呢？（答案保留小数点后两位有效数字）

然后，你还可以把小数转换为百分数。百分数就是分母是100的分数。你需要做的，就是将一个小数中的小数点向右移动两位，并在末尾加上百分号。

2. 将1/6表示成百分数是多少呢？

你可以自己决定用哪种方式表示概率，但知道概率的这三种表达方式对应用概率是有益的。

你也可以记住一些分数和百分数之间的简单换算关系。这里给出几个最简单的换算公式：

$$1/4 = 25\% = 四分之一$$
$$1/3 = 33.33\% = 三分之一$$
$$1/2 = 50\% = 二分之一$$
$$2/3 = 66.67\% = 三分之二$$
$$3/4 = 75\% = 四分之三$$

现在可以做一些有关概率的练习了。

3. 在一次旋转轮盘游戏中，如果轮盘上有从1到6几个数字，那么指针指向大于或等于3的数字的概率是多少？用分数表示。如果能够化简，化简后的结果是多少？

4. 这个概率用小数表示的话，是多少？

5. 这个概率用百分数表示的话，是多少？

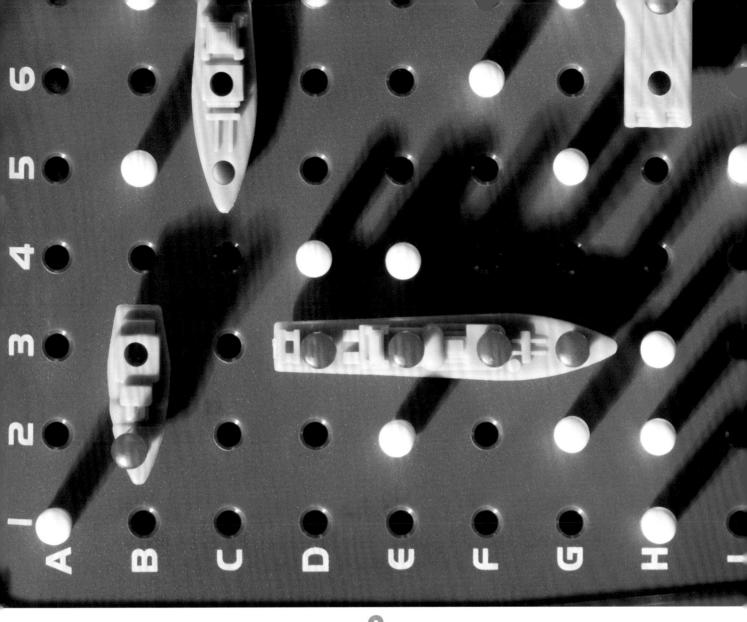

3
战舰游戏

梅森家里有一套战舰游戏，他正在教他的弟弟迈克尔如何玩这个游戏。梅森将自己的战舰放在游戏板的网格上，告诉迈克尔同样操作。然后梅森说，游戏双方将利用网格坐标来猜测对手战舰的位置并击沉它。

网格是10×10的，也就是说，网格有10个横孔和10个竖孔。横孔称为行，竖孔称为列。在下页你将看到梅森的战舰在游戏板上的位置，他的战舰用X表示。

	1	2	3	4	5	6	7	8	9	10
A								X	X	X
B		X		X						
C		X		X			X			
D		X		X			X			
E				X						
F										
G	X	X	X							
H						X	X	X	X	X
I										
J										

梅森让迈克尔先猜。梅森告诉迈克尔，他必须明确给出网格上一个孔的行和列，其中，行用字母表示，列用数字表示。

迈克尔猜到：G5。

1. 迈克尔说的G5是什么意思呢？梅森应该在什么位置放置一个钉子以表示迈克尔的猜测？迈克尔击中梅森的战舰了吗？

现在轮到梅森猜了。他猜B7并击中迈克尔的战舰了！他需要继续尝试猜测，并找到迈克尔其他位置的战舰并击沉它。
梅森下次击中的可能性有四种。

2. 梅森的最佳猜测是什么？为什么？

梅森知道他下次击中的可能性很大，但也可能猜错。
梅森猜测中有一次击中的概率是多少？请用分数表示。首先应该知道梅森一共有多少种选择，这个数就是分母，而他击中的次数就是分子。

3. 梅森下次击中的概率是多少？

4

"大富翁" 游戏

星期天晚上是梅森和家人的游戏之夜，这天晚上家庭成员聚在一起玩游戏。有时他们尝试新游戏，有时他们玩喜欢的老游戏。

梅森的爸爸总是想玩"大富翁"游戏，而梅森却不喜欢，或者至少他认为自己不喜欢。这次，梅森的爸爸坚持大家一起玩这个游戏，他认为如果玩一次，梅森或许就会喜欢这个游戏。梅森的爸爸还认为这个游戏能使梅森学会金钱管理技巧，因为有时梅森花费太多，有时又不知道钱放在哪儿了。

他们摆列好开始玩游戏。不久，梅森就进入了状态。他喜欢数学，他发现"大富翁"游戏与数学有很大的关系，特别是在钱增加的时候。很快，他就喜欢上了这个游戏，并想每个游戏之夜都玩这个游戏。

在这个游戏中，梅森扮演银行家。开始时，他需要给每位参与者1500美元。共有五位参与者——他的父母、弟弟迈克尔和姐姐劳拉。

1. 他一共需要拿出多少钱？

他不能正好给每个人一张1000美元和一张500美元的钞票，这个游戏甚至不允许使用1000美元的钞票。

梅森的妈妈给了他一些游戏中使用的钞票，包括：

10张500美元
20张100美元
5张50美元
5张20美元
10张10美元
5张5美元
25张1美元

梅森需要用每种面额的钞票的数量除以参与的人数，计算出他们各需要多少张。

2. 他应该给每个人每种面额的钞票多少张？

梅森和家人玩了一会儿。最后，在游戏结束时，梅森只剩下55美元。他想买价值180美元的圣杰姆斯地产。

下面是梅森拥有的街道，其他人如果从这里通过，需支付给他相应的过路费：

（橙色）田纳西州大道，14美元
（绿色）和平大道，26美元
（浅蓝色，一所房子）东方大道，30美元
（浅蓝色，两所房子）佛蒙特大道，90美元
（浅蓝色，一所房子）康涅狄格州大道，40美元

3. 梅森购买圣杰姆斯地产还差多少钱？

4. 是否存在两个地方，使得其他人支付给梅森的过路费足够他购买圣杰姆斯地产。

5
拼字游戏

　　天晚上，梅森和家人在玩拼字游戏。梅森的姐姐劳拉特别擅长玩拼字游戏，因为她学会了很多单词，她喜欢阅读而且几乎过目不忘。

　　梅森也擅长拼字游戏。尽管他的单词量不如劳拉，但他擅长将数学知识运用于游戏之中。游戏板上有很多空，标着"双倍字母得分""三倍字母得分"，等等。梅森知道，利用这些空格可以使单词获得更多的分数。实际上，他喜欢将最大分值的字母填入"三倍字母得分"的空格中。

　　具有分值的字母的数量是由字母在英语中的使用频率决定的，这就要用到数学了。下页我们将看到梅森如何赢得高分。

　　拼字游戏中每个字母的分值从1分到10分，在英语中出现频率越低的字母，其分值越高，因为这些字母也很难在拼字板上被用到。下面给出一些例子：

A的分值是1分
W的分值是4分
Q的分值是10分
P的分值是3分
G的分值是2分
K的分值是5分
X的分值是8分

1. 你希望这些字母在英语中以什么顺序排列，从最常用到最不常用？

现在我们关注分值。下面是游戏进行到一半时梅森一家游戏板上的情况：

双倍字母得分				三倍单词得分
		三倍字母得分		
U	S	U	A 双倍字母得分	L
双倍单词得分				双倍字母得分
	双倍单词得分			
		三倍字母得分		

梅森想尽可能赢得最大分值。他想用单词"QUIZ"，因为这个单词中有两个10分的字母，Q和Z的分值都是10分，U和I是1分。

游戏板上有单词"USUAL"，梅森可以利用其中的一个"U"来组成单词"QUIZ"。

梅森利用哪个"U"能赢得更多分值？请填入游戏板。

2. 梅森填写的这个词，能得多少分？

6
国际跳棋: 游戏活动区域

梅森如此喜欢玩游戏, 以至于他想自己设计游戏盘。他知道, 设计一个游戏盘是很困难的, 于是他从设计简单的游戏盘——国际跳棋棋盘开始。

国际跳棋棋盘是一个由红色和黑色方块相间的大方板, 红色和黑色方块的大小相同。为了做些改进并使游戏更有趣, 梅森想设计一款由绿色和紫色方块组成的游戏盘。他也想将新的游戏盘设计得更大一些, 尺寸是国际跳棋棋盘的两倍。

他需要并清楚如何设计游戏板。为此，他用到了长、宽、面积、表面尺寸。使用这些数据，梅森可以做一个专业的游戏板。

国际跳棋棋盘是正方形的，正方形是长和宽相等的长方形。标准的国际跳棋棋盘有1英尺（或12英寸）见方。

梅森想将他的游戏盘的长、宽设计成国际跳棋棋盘长、宽的两倍。

1. 梅森的游戏盘的每条边有多少英寸？

标准的国际跳棋棋盘横向和竖向各有8个方块，共计64个方块。棋盘的长和宽都被分成了8个部分。

2. 标准的国际跳棋棋盘的每个方块长度是多少？

3. 梅森设计的新游戏盘呢？新游戏盘的面积是标准游戏盘的两倍吗？

现在我们计算标准游戏盘的面积。长方形的面积公式是：

$$面积 = 长 × 宽$$

长方形的长称为水平线，长方形的宽称为垂直线（把长方形放在一张纸上看）。因为正方形的长和宽相等，所以面积公式变为

$$面积 = 宽 × 宽（或长 × 长）$$

4. 标准的国际跳棋棋盘的面积是多少平方英尺？多少平方英寸？

5. 梅森设计的新游戏盘的面积是多少平方英尺？多少平方英寸？

6. 新游戏盘的面积是标准游戏盘面积的两倍吗？如果不是，是多少倍呢？

你将会发现，如果长方形的长和宽增大到两倍，则面积是原来面积的4倍。

7
石头、剪子、布

梅森和弟弟、妹妹每天放学回家都要做一些家务。一天晚上，他们在晚餐后负责清理卫生。一个人洗盘子，一个人烘干，一个人将烘干的盘子收起来。

劳拉上次刷了盘子，她这次不用再刷盘子了。梅森和迈克尔也不想刷盘子，因为刷盘子看起来是最累的一项工作。梅森建议他俩玩石头、剪子、布的游戏，输的人刷盘子。两个男孩都喜欢玩这个游戏，因此他们要在七局游戏中好好发挥，看看谁能在七局游戏中赢的最多。

下页表格中给出了石头、剪子、布在实际中出现的结果。

根据概率知识，可以知道游戏者选择石头、剪子、布的概率都是1/3。下面看一下梅森和迈克尔在游戏中的实际情况：

梅森：

轮	结果
1	石头
2	石头
3	剪子
4	布
5	石头
6	剪子
7	石头

迈克尔：

轮	结果
1	布
2	石头
3	布
4	石头
5	剪子
6	剪子
7	布

梅森选择石头的概率是4/7，选择布的概率是1/7，选择剪子的概率是2/7。

如果迈克尔是一台计算机，他选择每一种的概率都是1/3。但他是一个自然人，由于个人喜好，他更喜欢选择出石头。或许他选择石头赢的次数更多，或许他认为石头比布或剪子更厉害。

1. 梅森选择哪种的概率小于1/3？哪种大于1/3？

2. 迈克尔选择石头、剪子、布的概率各是多少？

3. 迈克尔选择哪种的概率小于1/3？哪种大于1/3？

现在算出该谁刷盘子。石头胜剪子，剪子胜布，布胜石头。补充填写下表：

轮	胜者
1,	迈克尔
2,	平
3,	
4,	
5,	
6,	
7,	

4. 谁赢了？他赢的概率是多少？

8

纸牌游戏：战争

放 学后，梅森去他的朋友迪亚哥家玩。迪亚哥特别喜欢纸牌游戏，他俩总是玩很长时间。

今天，梅森和迪亚哥玩战争游戏。他们先从一种简单的玩法开始。将纸牌平均分成两份，每次翻一张，牌大者赢得这两张牌。如果两张牌大小一样，他们就要拿出4张牌，3张正面朝下，1张正面朝上，正面朝上的牌大的人赢得所有的牌。赢牌最多的人获胜。

然后他们开始玩一种复杂的玩法。他们需要用心算练习在学校学习的乘法，出于这个目的，因此他们发明了这个游戏。他们不是一次只翻一张牌了，而是一次翻两张，并将这两张牌的数值相乘。乘积大的人赢得所有的牌。下页我们将会看到他们是怎么玩的。

在很多纸牌游戏中，我们也需要记住花牌的分值，J是11，Q是12，K是13，A是1.

战争游戏简单表示如下：

梅森	迪亚哥
7	3
Q	10
5	9
3	4
J	K

1.　游戏结束后谁赢了？在这5局游戏中每人各赢了多少张纸牌？

然后他们开始玩乘法战争游戏。他们翻出两张牌并将数值相乘，乘积大者赢得所有纸牌。游戏情况如下：

梅森	迪亚哥
6 × 9 =	Q × 2 =
K × A =	9 × 4 =
3 × 5 =	A × 5 =
7 × J =	10 × 8 =
2 × 7 =	K × J =

2.　第一轮谁赢了？

3.　游戏结束后谁赢了？

4.　游戏的胜方比败方多了多少张纸牌？

9

纸牌游戏：概率

纸牌游戏也给了梅森和迪亚哥练习概率知识的机会。纸牌游戏和概率密切相关，所以，当你玩纸牌游戏时，你就会发现相关的概率问题很容易理解。纸牌有不同的颜色、不同的花色和不同的数字。几乎你玩的每种纸牌游戏在某种程度上都最终变成处理概率问题。

谁能想到纸牌游戏能训练我们的数学能力？下面我们将会看到，纸牌游戏是如何训练我们的数学能力的。

一副标准的纸牌有52张，分为4个花色：红桃、方片、梅花和黑桃。红桃和方片是红色的，梅花和黑桃是黑色的。

1. 抽中一张红色牌的概率是多少？约分后是多少？

2. 抽中一张花牌（J，Q，K）的概率是多少？约分后是多少？

观察每一个类型的牌，每个数字的牌都有4张（4张5，4张K，等等）。

3. 抽中一张A的概率是多少？

如果我们抽中后不放回，那么连续抽中4张A的概率是多少？你可能会想，"不会很高"，但这个概率究竟有多么小呢？这个问题有点复杂。

你需要分别计算抽中每一张A的概率，然后将这些概率相乘。

你已经知道了抽中第一张A的概率，现在，抽中第二张A的概率是多少？每抽出一张牌，你就要记住整副牌中少了一张，所以再次抽中这个数字的牌的可能性就小了一点。就是说，你已经抽中了一张A，所以整副牌中的A也少了。

4. 抽中第二张A的概率是多少？

5. 抽中第三张和第四张A的概率分别是多少？

现在可以将这四个概率相乘了。

6. 这些概率的分数形式是什么？百分数形式呢？

10
数　　独

　　一天放学后，梅森看到爸爸在报纸上玩填字游戏。其实不是填字游戏，准确地说，应该是填数字游戏。梅森的爸爸告诉他这是数独，并告诉梅森游戏规则。

　　这个游戏由9×9的方格网组成，这个大方格被粗线分割成9个小方格，每个小方格由3×3的更小的方格组成（看下页图）。这个方格中已经给出一些数字，游戏的目标是填满剩余的方格。你只能用数字1到9这九个数字。在9组粗线组成的小方格内，数字不能重复；任意一行或一列的数字也不能重复。

　　梅森明白，玩家必须擅长加法、模型和逻辑。观察下页的数独和问题，然后寻找答案。

			8					
4				1	5		3	
	2	9		4		5	1	8
	4					1	2	
			6		2			
	3	2					9	
6	9	3		5		8	7	
	5		4	8				1
					3			

试着在第三行填入数字，因为第三行中已经给出了很多数字。

你需要使用试错法。可能你找到了一个数字而解决了一组问题；也可能你发现这个数字并不正确，那就不得不换掉这个数字，重新开始。

在第三行的第一个小方格内，你填入的数字不能是1,2,4,5,8,9，也可以排除6，因为6已经在第一列中。

接下来，你可以尝试填写第七行的第四个小方格。你填入的数字不能是3，5，6，7，8，9，因为这些数字已经在这一行中。也不能是4，因为4出现在了这组小方格内，也出现在同一列中。

尝试填满一个数独，完成第一个数独时可能会感到很难，但不要沮丧。如果能够坚持下去，就能对数独背后蕴含的逻辑多一点理解。

11
保　龄　球

梅森的朋友伯瑞娅邀请他玩保龄球，她知道他喜欢玩游戏，所以认为他可能也喜欢玩保龄球。她猜对了——梅森喜欢玩保龄球，并且非常喜欢它的竞技性。

每当玩保龄球的时候，他喜欢自己计算得分。当然计算机会给出得分，但他更喜欢自己确认所赢得的分数。他也计算每个人的得分，而不仅仅自己的。

伯瑞娅的朋友杰罗姆也在保龄球馆，他以前从没玩过保龄球，于是梅森向他解释保龄球的规则：

每位选手在每一轮中有两次投球机会。

每一局游戏有10轮。

如果一次击倒了全部保龄球瓶，即为全中，得10分，该轮得分为10分加上下轮两球的所得分。当你投了一次全中，那么在下一轮结束之前，并不能计算出这轮的得分。

如果用两次机会击倒全部保龄球瓶，即为补中，得10分，该轮得分为10分加上下一轮第一个球的得分。

如果在一轮中第二个球也没有把保龄球瓶全部击倒，那么这一轮的得分就是击倒的所有保龄球瓶数。

如果在最后一轮中击倒全部保龄球瓶，则额外奖励一个球。

下页给出如何计算一局游戏的得分。

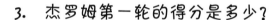		1	2	3	4	5	6	7	8	9	10	T	

佰瑞娅首先投球。第一个球她击倒了2个瓶，第二个球补中。于是只能等到她投完下一轮之后才能计算出得分。在下一轮，她投了一个全中。

1. 佰瑞娅第一轮的得分是多少？

梅森第一轮就投出了一个全中，他也不能算出他的得分，因为还要加上下一轮两个球的得分。在下一轮，第一个球击倒1个瓶，第二个球击倒7个。

2. 梅森第一轮的得分是多少？

杰罗姆在第一轮中，第一个球击倒2个瓶，第二个球击倒5个。

3. 杰罗姆第一轮的得分是多少？

下面是这一局游戏中各轮的得分，将这些信息填入得分卡，全中记为"X"，补中记为"/"。

第1轮：　佰瑞娅——2，补中；　梅森——全中；　　杰罗姆——2，5
第2轮：　佰瑞娅——全中；　　梅森——1，7；　　杰罗姆——7，补中
第3轮：　佰瑞娅——3，补中；　梅森——6，3；　　杰罗姆——4，2
第4轮：　佰瑞娅——4，0；　　梅森——全中；　　杰罗姆——8，0
第5轮：　佰瑞娅——7，2；　　梅森——2，补中；　杰罗姆——8，1
第6轮：　佰瑞娅——0，4；　　梅森——4，3；　　杰罗姆——2，5
第7轮：　佰瑞娅——8，1；　　梅森——4，5；　　杰罗姆——1，8
第8轮：　佰瑞娅——0，3；　　梅森——2，0；　　杰罗姆——4，4
第9轮：　佰瑞娅——9，补中；　梅森——0，9；　　杰罗姆——4，0
第10轮：佰瑞娅——7，1；　　梅森——2，7；　　杰罗姆——9，补中，7

4. 谁赢了这一局？他或她得了多少分？

5. 另外两位选手各得多少分？

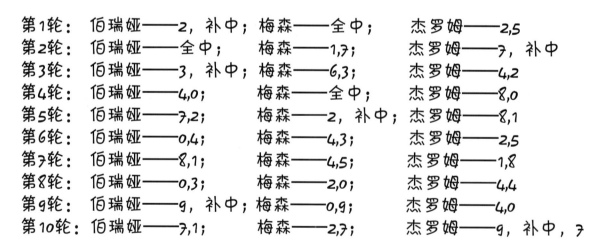

22

12

台　　球

梅森喜欢的另一项游戏是台球，他刚刚开始跟姐姐劳拉学习台球。劳拉台球打得很好，而且懂得很多技巧。

一天，劳拉和梅森在家里的地下室练习打台球，她决定传授梅森一些技巧。她告诉梅森，台球与几何和角度相关，角度是两条相交直线所构成的空间，用度来表示。

如果以一定角度击打主球去碰撞目标球，目标球将以同样的角度弹开。例如，如果以180度直接击球，那么它也将以180度的角度弹开。如果主球、目标球和球袋在一条直线上，那么这将是很好的一次击球。

打台球时，不一定能精确计算出击球的角度，但可以估计一下。下页给出一些角度，并说明如何利用角度成为一名出色的台球选手！

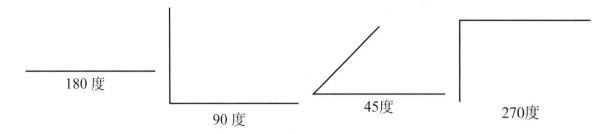

180 度   90 度   45 度   270 度

梅森从容易击中的球开始。目标球和球台一侧中袋在一条直线上，而且主球也在这条直线上。

如图A，白球是主球，黑球是需要击入袋的目标球。你甚至可以从主球到目标球再到球袋画一条直线，以便对角度有更直观的感觉。

1. 如果要使主球、目标球和球袋在一条直线上，他的击球角度是多少？

你也可以击打主球，主球碰撞目标球，使其撞击球台并以某角度滚入袋中。在图B中，虚线表示应该从什么角度撞击右边的球，以使其反弹入袋。

2. 这个角度约是多少？

第三种得分的方法是以一个角度击打主球，使其碰撞球台边沿后弹回，然后碰撞目标球，使其入袋。

图B中左侧的球的运动轨迹就是这样的例子。

3. 这个角度约是多少？

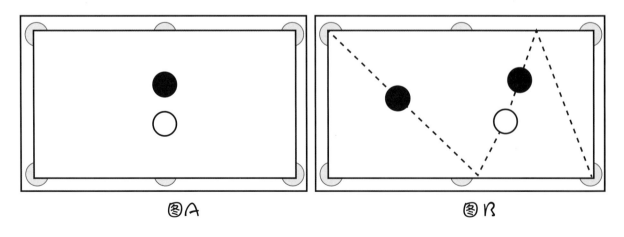

图A   图B

24

13

电子游戏：线性代数

梅森也喜欢玩电子游戏，在不玩桌上游戏或纸上（或计算机上）游戏时，他就玩电子游戏。

梅森最喜欢的一款电子游戏是很简单的。它需要操作一艘绕着小行星飞行的飞船去撞击小行星，如果飞船没有以足够的速度撞毁小行星，飞船将坠毁而游戏结束。

这款电子游戏用到线性代数，涉及其中称为矢量的数学知识。矢量是有方向的单点，当这个单点移动时，沿着其移动的方向构成线。

在梅森的电子游戏中，移动飞船是一个矢量，小行星也是。下页的内容将会帮助我们更好地理解矢量。

大圆点代表飞船，如果它沿着箭头所示方向移动，其矢量是什么？

飞船向左移动了3个格，在线性代数中，这意味着它的移动距离是-3（负3）；向右移动则为正数。飞船也向上移动了2个格，在线性代数中，这意味着它移动距离为2，向下移动为负数。

这个矢量的轨迹记作（-3，2）。

水平移动在前，垂直移动在后。

1. 如果飞船向右移动1个格向上移动4个格，那么这个矢量是什么？请在网格上画出来。

2. 如果飞船向左移动2个格向下移动1个格呢？请在网格上画出来。

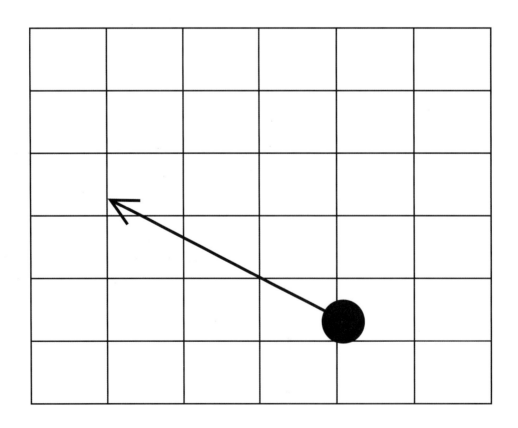

14

更多电子游戏数学：科学符号

在飞船电子游戏中，梅森努力获得更多得分，由于他已经玩了很长时间，他不得不和非常大的数字打交道。每得到1000点，他就能得到1颗宝石；得到10颗宝石后，他就能得到一颗行星。

这个游戏的最终目标是收集整个太阳系中所有有价值的行星，在这款电子游戏中，是100颗行星。梅森想知道这需要获得多少点。他可以用数学方法计算出来——请看下页。

如果想要计算出游戏通关需要多少点，首先，计算得到一颗行星需要多少点。

1颗行星=10颗宝石，   1颗宝石=1000点

用1颗宝石需要的点数乘以1颗行星需要的宝石数，得到

1. $10 \times 1000 =$

现在，用游戏通关所需要的行星数（即100）乘以每颗行星的点数。

2.   赢得游戏需要多少点呢？

注意到，在答案中有非常多的0，数学上给出了表示非常多的0的很大的数字的简便方法，称为科学计数法。例如：

$$10^1 = 10$$
$$10^2 = 100$$
$$10^3 = 1000$$

现在，填写下面的空白：

$$10^4 =$$
$$10^5 =$$
$$10^6 =$$
$$10^7 =$$
$$10^8 =$$

3.   如何用科学计数法表示游戏通关所需要赢得的宝石数？

用科学计数法可以很方便地进行乘法和除法运算，对于乘法，只要指数相加即可；对于除法，则是指数相减。

游戏通关所需要赢得的点数的公式即为

$$10^1 \times 10^3 \times 10^2$$

因为是做乘法，所以指数相加。

4.   所得的结果与前面一样吗？

15
小　　结

梅森从未认识到游戏和数学的联系有多紧密，二者他都喜欢，所以二者联系如此紧密对他是很有意义的。

我们能够看到数学在游戏中是多么有用。数学能帮助我们更好地理解游戏，甚至在游戏中获胜。看看你记得多少梅森喜欢的游戏。

1. 掷六面骰子时，掷出的数字是1或者大于4的概率是多少？

这个分数可以简化吗？如果可以，简化后是多少？

这个概率的小数形式是什么？

百分数形式呢？

2. 在大富翁游戏中，计算游戏结束时钱的总数。你的对手有2178美元，你有2张500美元，6张100美元，9张50美元，7张20美元，2张10美元和11张1美元的钞票。

你获胜了吗？

如果是，你赢了多少钱；如果否，你还需要赢多少钱？

3. 大小14英寸×22英寸的矩形游戏盘的面积是多少？

4. 在乘法战争游戏时，你放下一张Q和一张8，对手放下一张J和一张9。谁获胜了？

5. 在52张牌中抽出一张红A的概率是多少？

这个概率约分后是多少？

6. 在打保龄球时，你第一轮补中，接着在下一轮分别击中了8个瓶和1个瓶；你的朋友第一轮全中，下一轮分别击中了2个瓶和4个瓶。

在第一轮中谁得分高？

7. 在电子游戏中，矢量箭头向左移动4个格，向上移动7个格，这个矢量是多少？

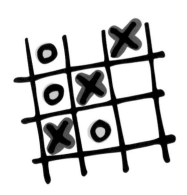

参考答案

1.

1. 6
2. 1
3. 1/2; 也可以说概率是一半
4. 1/12小, 因为有更多的结果, 任何一个数出现的机会就会变化。1/20更小, 因为有更多的结果
5. 6/20, 或3/10

2.

1. 0.17
2. 17%
3. 4/6, 2/3
4. 0.67
5. 66.67%

3.

1. 第G行, 第5行的交点; 没有, 他没有击中战舰
2. B6, B8, A7, C7; 它们都邻近他的正确答案
3. 1/4

4.

1. 1500美元 × 5 = 7500美元
2. 2张500美元, 4张100美元, 1张50美元, 1张20美元, 2张10美元, 1张5美元和5张1美元
3. 125美元
4. 存在——佛蒙特大道和唐涅狄格州大道

5.

1. A, G, P, W, K, X, Q
2. 44分

6.

1. 24英寸
2. 1 1/2 英寸
3. 3 英寸; 不是
4. 1 平方英尺, 144 平方英寸
5. 4 平方英尺, 576 平方英寸
6. 不是, 是原来的4倍

7.

1. 小: 1/7, 2/7; 大: 4/7
2. 石头 = 2/7, 布 = 3/7, 剪子 = 2/7
3. 小: 2/7, 大: 3/7

轮,	胜者
1,	迈克尔
2,	平
3,	梅森
4,	梅森
5,	梅森
6,	平
7,	迈克尔

4. 梅森赢, 比率是43%

8.

1. 迪亚哥赢, 在这几轮比赛中, 迪亚哥有6张牌, 梅森只有4张
2. 梅森
3. 迪亚哥
4. 迪亚哥比梅森多4张牌

梅森	迪亚哥
6 × 9 = 54	Q × 2 = 24
K × A = 13	9 × 4 = 36
3 × 5 = 15	A × 5 = 5
7 × J = 77	10 × 8 = 80
2 × 7 = 14	K × J = 143

9.

1. 26/52, 1/2
2. 12/52, 3/13
3. 4/52 (或 1/13)
4. 3/51
5. 第三张: 2/50, 第四张: 1/49
6. 24/6497400, 0.000369378%

10.

3	1	5	8	2	7	9	4	6
4	6	8	9	1	5	7	3	2
7	2	9	3	4	6	5	1	8
9	4	6	5	3	8	1	2	7
5	7	1	6	9	2	4	8	3
8	3	2	1	7	4	6	9	5
6	9	3	1	5	1	8	7	4
2	5	7	4	8	9	3	6	1
1	8	4	7	6	3	2	5	9

11.

1. $10 + 10 = 20$
2. $10 + 8 = 18$
3. 7
4. 伯瑞娅赢，她得了108分
5. 梅森得了105分，杰罗姆得了89分

12.

1. 180 度
2. 45 度
3. 90 度

13.

1. (1, 4)
2. (−2, −1)

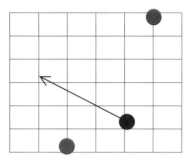

14.

1. 10000
2. 1000000

$10^4 = 10000$
$10^5 = 100000$
$10^6 = 1\,000000$
$10^7 = 10000000$
$10^8 = 100000000$

3. 10^6
4. 是的

15.

1. 3/6; 是的, 1/2; 0.50; 50%
2. 是的; 你赢了43美元
3. 14 × 22 = 308 平方英寸
4. 你对手赢了
5. 2/52; 1/26
6. 你得到了最高分: 18
7. (−4, 7)

INTRODUCTION

How would you define math? It's not as easy as you might think. We know math has to do with numbers. We often think of it as a part, if not the basis, for the sciences, especially natural science, engineering, and medicine. When we think of math, most of us imagine equations and blackboards, formulas and textbooks.

But math is actually far bigger than that. Think about examples like Polykleitos, the fifth-century Greek sculptor, who used math to sculpt the "perfect" male nude. Or remember Leonardo da Vinci? He used geometry—what he called "golden rectangles," rectangles whose dimensions were visually pleasing—to create his famous *Mona Lisa*.

Math and art? Yes, exactly! Mathematics is essential to disciplines as diverse as medicine and the fine arts. Counting, calculation, measurement, and the study of shapes and the motions of physical objects: all these are woven into music and games, science and architecture. In fact, math developed out of everyday necessity, as a way to talk about the world around us. Math gives us a way to perceive the real world—and then allows us to manipulate the world in practical ways.

For example, as soon as two people come together to build something, they need a language to talk about the materials they'll be working with and the object that they would like to build. Imagine trying to build something—anything—without a ruler, without any way of telling someone else a measurement, or even without being able to communicate what the thing will look like when it's done!

The truth is: We use math every day, even when we don't realize that we are. We use it when we go shopping, when we play sports, when we look at the clock, when we travel, when we run a business, and even when we cook. Whether we realize it or not, we use it in countless other ordinary activities as well. Math is pretty much a 24/7 activity!

And yet lots of us think we hate math. We imagine math as the practice of dusty, old college professors writing out calculations endlessly. We have this idea in our heads that math has nothing to do with real life, and we tell ourselves that it's something we don't need to worry about outside of math class, out there in the real world.

But here's the reality: Math helps us do better in many areas of life. Adults who don't understand basic math applications run into lots of problems. The Federal Reserve, for example, found that people who went bankrupt had an average of one and a half times more debt than their income—in other words, if they were making $24,000 per year, they had an average debt of $36,000. There's a basic subtraction problem there that should have told them they were in trouble long before they had to file for bankruptcy!

As an adult, your career—whatever it is—will depend in part on your ability to calculate mathematically. Without math skills, you won't be able to become a scientist or a nurse, an engineer or a computer specialist. You won't be able to get a business degree—or work as a waitress, a construction worker, or at a checkout counter.

Every kind of sport requires math too. From scoring to strategy, you need to understand math—so whether you want to watch a football game on television or become a first-class athlete yourself, math skills will improve your experience.

And then there's the world of computers. All businesses today—from farmers to factories, from restaurants to hair salons—have at least one computer. Gigabytes, data, spreadsheets, and programming all require math comprehension. Sure, there are a lot of automated math functions you can use on your computer, but you need to be able to understand how to use them, and you need to be able to understand the results.

This kind of math is a skill we realize we need only when we are in a situation where we are required to do a quick calculation. Then we sometimes end up scratching our heads, not quite sure how to apply the math we learned in school to the real-life scenario. The books in this series will give you practice applying math to real-life situations, so that you can be ahead of the game. They'll get you started—but to learn more, you'll have to pay attention in math class and do your homework. There's no way around that.

But for the rest of your life—pretty much 24/7—you'll be glad you did!

1
DICE AND PROBABILITY

Mason loves to play games. He plays them whenever he can—at school, at home on the weekends, at friends' houses after school, on the bus. Mason will play just about every kind of game he can find.

Mason also really likes math, which is his favorite subject in school. One day in school, he learns about probability, which is the chance that something will happen. His teacher was explaining about probabilities as expressed in fractions.

Later, when he is playing around with some dice, he realizes that dice actually have a lot to do with probability. Mason has just discovered that most games, especially ones with dice, are based at least partly on probability. Check out the next page to see how probability relates to dice.

Mason is playing with a regular, six-sided die. Each side is labeled with one to six dots. He wants to know what the possibility of rolling a 3 is.

First he thinks about how many outcomes—or possibilities—there are when rolling a dice. The outcome is the denominator (the bottom number) of the probability fraction.

1. When he throws the die, how many outcomes are there?

To get the numerator of the fraction, you need to think about how many ways there are to get an event to happen. In this case, you want to think about how many ways Mason could roll a 3.

2. How many ways of rolling a 3 are there?

Now put those two numbers together into a probability fraction: ⅙. You could also say that Mason has a one in six chance of rolling a 3.

You can also calculate the probabilities of rolling more than one number on a die. The probability of rolling an even number, for example, is ⅗. There are three ways to roll an even number, and six outcomes.

If you can reduce the fraction, you should. That way, you might be able to understand what the probability really means a little better.

3. What is ⅗ reduced? Can you think of a different way of saying that probability?

Mason also has some crazy dice at home. One has twelve sides, and he even has a twenty-sided die. You can do some more complicated probabilities with these dice.

The probability of rolling a 3 on a twelve-sided die is 1⁄12 and on a twenty-sided die, it is 1⁄20.

4. Is 1⁄12 a bigger or smaller probability than 1/6? Why? What about a 1⁄20 probability?

5. On the twenty-sided die, what is the probability that you roll a number that is divisible by 3? You may want to list all the numbers and then count them up first.

2
SPINNERS: MORE ON PROBABILITY

Many of the board games Mason plays have spinners. To move forward on the board, players spin an arrow that spins around in a circle and stops on one particular number. Just like dice, spinners rely on probability. If a spinner has six spaces, the probability of spinning a 3 is just the same as rolling a 3 on a six-sided die. The number of outcomes are the same, and the chance of getting a 3 is the same.

However, you can start to think about probability in different ways once you know the basics. Probabilities can be fractions, but they can also be decimals and percents.

You already know you can say that the probability of spinning a 3 is ⅙. The next step is to figure out how to **convert** that fraction to a decimal.

To convert a fraction to a decimal, simply divide the numerator (top number) by the denominator (bottom number).

1. What is a probability of ⅙ in decimal **numerals**? (Round your answer it to the nearest hundredth.)

Next, you can convert the decimal into a percent. Percents are parts out of 100. All you need to do is move the decimal point in a decimal over two places to the right and add a percent sign on the end.

2. What would ⅙ be as a percent?

It is up to you how you think about probabilities, but it is helpful to know all three forms.

You can even memorize some shortcuts between fractions and percents. Here are a few of the most common:

$$¼ = 25\% = \text{one-quarter}$$

$$\frac{1}{3} = 33.33\% = \text{one-third}$$
$$\frac{1}{2} = 50\% = \text{one-half}$$
$$\frac{2}{3} = 66.67\% = \text{two-thirds}$$
$$\frac{3}{4} = 75\% = \text{three-quarters}$$

Now do some more probability practice:

3. What is the probability, as a fraction, of spinning a number greater or equal to 3 on a game board with a spinner numbered 1 through 6? If it can be reduced, what does it reduce to?

4. What is that same probability as a decimal?

5. And what about the same probability as a percentage?

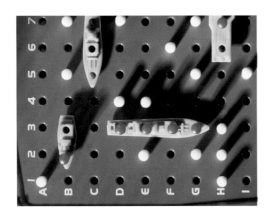

3
BATTLESHIP

Mason has the game Battleship at home, which he is teaching his younger brother Michael how to play. Mason places his own ships on the board's grid. He tells Michael to do the same thing. Then Mason explains they are going to use the grid coordinates to guess where the other person's ships are and sink them.

Each grid is 10 by 10. In other words, it has ten spaces across and ten spaces down. The spaces across are called rows, and the spaces down are called columns. You'll see Mason's Battleship board on the next page. His ships are marked with Xs.

	1	2	3	4	5	6	7	8	9	10
A								X	X	X
B		X		X						
C		X		X			X			
D		X		X			X			
E				X						
F										
G	X	X	X							
H						X	X	X	X	X
I										
J										

Mason lets Michael guess first. He tells him he has to identify a space on the grid by both its row and column. The rows are letters and the columns are numbers.

Michael makes a guess. He says, "G5."

1. What did Michael mean? Where should Mason put the peg to show Michael's guess? Did he get a hit?

Now Mason guesses. He guesses B7 and gets a hit! He has to take another guess and try and find the rest of the ship to sink it all.
There are four guesses Mason could make next to get another hit.

2. What are the best guesses he could make? Why?

Mason knows he has a good chance of getting another hit, but he might miss with his guess too.
What is the probability one of those guesses will hit a ship? You can show the probability as a fraction. First ask yourself how many choices he has. That will be the denominator. The numerator will be how many guesses he has.

3. What is the probability of getting another hit?

4
MONOPOLY

Mason and his family have game nights on Sunday nights. Everyone gets together and plays a game. Sometimes they try out new games, and sometimes they play old favorites.

Mason's dad always wants to play Monopoly. Mason hates Monopoly, though, or at least thinks he hates it. This time, his dad insists they all play. He thinks Mason really will like it if he gives it a chance. His dad also thinks it will teach him better money **management** skills, because sometimes Mason spends too much money, or loses track of where he put his cash.

They open the box and spread everything out and start playing. Soon, Mason is really getting into the game. He likes math, and he's finding that Monopoly is all about math, especially adding money. Pretty soon, he has changed his mind about the game and wants to play it every game night!

Mason is acting as the banker for this game. At the beginning, he has to give each person $1500. There are five people playing—his parents, his brother Michael, and his sister Lora.

1. How much money does he need to hand out in all?

He can't just give everyone a thousand-dollar bill and a five-hundred-dollar bill. The game doesn't even have a thousand-dollar bill.

Mason's mom gives him some bills to hand out. She gives him:

>   ten $500 bills
>   twenty $100 bills
>   five $50 bills
>   five $20 bills

ten $10 bills
five $5 bills
twenty-five $1 bills.

He will need to divide the number of each of these kinds of bills by the number of players to figure out how many of each they need.

2. How many of each kind of bill should he give to each player?

Mason and his family play for a while. Eventually, he ends up spending all his money except for $55. He wants to buy St. James Place for $180.

Here is what he owns, and how much players have to pay him if they land on it:

(orange) Tennessee Avenue, $14
(green) Pacific Avenue, $26
(light blue, one house) Oriental Avenue, $30
(light blue, two houses) Vermont Avenue, $90
(light blue, one house) Connecticut Avenue, $40

3. How much money will Mason need to buy St. James Place?

4. Are there any combinations of two spaces other players can land on which will give Mason enough money to buy St. James Place?

5
SCRABBLE

Another night, Mason and his family play Scrabble. Mason's sister, Lora, is especially good at Scrabble because she knows so many words. She loves to read all the time, and remembers just about every word she reads.

Mason is also good at Scrabble, however. He isn't as good with words as Lora is, but he pays attention to the math on the board. The board has lots of spaces marked "Double Word Score," "Triple Letter Score," and more. He knows he can use those spaces to make his words worth a lot more. He really likes to use the letters that are worth the most on the "Triple Letter Score" spaces.

The amounts the letters are worth are based on how frequently the letters are used in the English language, which is even more math! See how Mason gets his high scores on the next page.

Each of the Scrabble letters are worth 1 to 10 points. The less frequently the letters appear in English, the more they are worth. That's because they're harder to use on the Scrabble board. Here are some examples:

A is worth 1 point
W is worth 4 points
Q is worth 10 points
P is worth 3 points
G is worth 2 points
K is worth 5 points
X is worth 8 points

1. In what order would you expect the letters to be used in English, from most often used, to least often?

Now you can focus on scoring. Here's what a section of Mason's family's board looks like during the middle of the game:

Double letter score				Triple word score
		Triple letter score		
U	S	U	A Double letter score	L
Double word score				Double letter score

	Double word score			
		Triple letter score		

Mason wants to get the most points possible. He wants to use the word "quiz," because it has two 10 point letters. Q and Z are worth 10 each, U is worth 1, and I is worth 1.

The word "usual" is already on the board. He can use either "u" to build the word "quiz."

Which "u" should Mason use for more points? Fill it in on the board.

2. How many points does he get for his word?

6
CHECKERS: AREA OF A GAME BOARD

Mason likes games so much that he wants to design his own game boards. He knows it might be tricky to build a board, so he starts out with a simple game board: checkers. A checkers board is made up of alternating red and black squares arranged into a larger square. Each red and black square is the exact same size. Mason's goal is to create a new board with green and purple squares, just for something different and to make the game more fun. He also wants to make it really big, twice the size of a normal checker board.

He'll need to figure out how to design the board. For that, he'll use length, width, and area, the size of a surface. By using these measurements he can make a professional-looking game board.

Checkerboards are squares. And squares are just rectangles where the length and width are exactly the same. A normal checkerboard is one foot (or 12 inches) long.

Mason wants to make his checkerboard twice as wide and twice as long as a normal board.

1. How long will each side be in inches?

A normal checkerboard also has 8 squares across and 8 squares down, for a total of 64 squares. The width and the length of the board are divided into 8 segments.

2. How big is each square in inches on a normal board?

3. What about on the bigger board? Are the new squares twice as big as the original ones?

Now find the area of the normal board. The equation for the area of a rectangle is:

$$area = length \times width$$

with the length being how long the rectangle is horizontally, and the width being how wide it is vertically (if you are looking at it on a piece of paper). Since a square has the same length and width, the area equation becomes:

$$area = width \times width \text{ (or length} \times \text{length)}$$

4. What is the area of a normal checkerboard in square feet? In square inches?

5. How big is the area of Mason's giant checkerboard in square feet? In square inches?

6. Is the new area twice as big as the normal area? If not, how many times bigger is it?

You have just discovered that as the sides of a rectangle double, the area quadruples!

7
ROCK, PAPER, SCISSORS

Mason and his brother and sister have to do chores every day after school. One night, they are in charge of cleaning up after dinner. One person has to wash the dishes, one person has to dry them, and one person has to put them away.

Lora washed the dishes last time, so she doesn't have to do it again. Neither Mason nor Michael wants to wash the dishes, because that seems like the most work. Mason suggests they play Rock, Paper, Scissors. The loser will have to wash the dishes. Both boys like playing games, so they make it best of seven, or whoever gets the most wins in seven games.

The next page reveals the real-life probability of Rock, Paper, Scissors, shown in charts. According to the rules of probability, you might expect there is a ⅓ chance a player will pick rock, a ⅓ chance a player will pick paper, and a ⅓ chance a player will pick scissors. Take a look at what actually happens in Mason and Michael's game:

Mason:

Turn	Outcome
1	Rock
2	Rock
3	Scissors
4	Paper
5	Rock
6	Scissors
7	Rock

Michael:

Turn	Outcome
1	Paper
2	Rock
3	Paper
4	Rock
5	Scissors
6	Scissors
7	Paper

The probability that Mason chose Rock was ⁴⁄₇. The probability he chose Paper was ¹⁄₇. And the probability he chose Scissors was ²⁄₇.

If Michael had been a computer, he would have chosen each 1/3 of the time. But because he is a human being, he chooses Rock more often than the other two, just because he likes to. Maybe he wins more often with Rock, or he thinks it's stronger than Paper or Scissors.

1. Which of Mason's probabilities was less than ⅓? Which was more than ⅓?

2. What were the probabilities for Michael?

3. Which probabilities were less than ⅓? Which were more than ⅓?

Now figure out who had to wash the dishes. Rock beats scissors, scissors beat paper, and paper beats rock. Fill out this chart:

Turn, Who Won
1, Michael
2, Tie
3,
4,
5,
6,
7,

4. Who won? What percent of the time did he win?

8
CARD GAMES: WAR

After school, Mason goes over to his friend Diego's house. Diego particularly likes card games, so the two friends spend a lot of time playing cards.

Today, Mason and Diego play War. They start out with the simple version. They divide the cards in half and flip one over at a time. The person with the higher card gets both cards. If the two cards are the same value, they have to put three more cards face down and another card face up. Whoever has the highest card flipped up wins all the cards. The player with the most cards wins.

Then they move on to a harder version. They need to practice multiplying in their heads for school, so they make a War game out of it. Instead of one card at a time, they each flip over two cards and multiply the values together. Whoever has the highest value takes all the cards. See exactly how they play on the next page.

In lots of card games, you need to remember that face cards have number values too. Jacks are 11, queens are 12, and kings are 13. Aces are 1.

The simple version of War looks like this:

Mason	Diego
7	3
queen	10
5	9
3	4
jack	king

1. Who is winning by the end of these rounds? How many cards does each boy have from these rounds?

Then they start playing Multiplication War. They put down two cards and multiply them together to see who gets to keep the cards. This is what their game looks like. Fill in the chart:

Mason	Diego
6 x 9 =	queen x 2 =
king x ace =	9 x 4 =
3 x 5 =	ace x 5 =
7 x jack =	10 x 8 =
2 x 7 =	king x jack =

2. Who wins the first round?

3. Who is winning by the end of these five rounds?

4. How many more cards does the winner of these rounds have than the person who is behind?

9
CARD GAMES: PROBABILITY

Cards also give Mason and Diego a chance to practice probability some more. Card games are all about probability, so you might find it easier to understand probability if you play around with cards. Cards have different colors, different suits, and different numbers. Almost every card game you play will end up dealing with probability of some kind.

Who knew card games could help you with math? Look on the next page to find out how. A standard card deck has 52 cards arranged into four suits: hearts, diamonds, clubs, and spades. Hearts and diamonds are red, while clubs and spades are black.

1. What is the probability of drawing a red card? What is the reduced probability?

2. What is the probability of drawing a face card (jack, queen, king)? What is the reduced probability?

Next, look at each type of card. There are four of each number card (four 5s, four kings, etc.).

3. What is the probability of drawing one ace?

What is the probability of drawing four aces in row if you don't replace them after you draw them? You're probably thinking, "not very high," but just how tiny is that probability? This question is a little more complicated.

You'll have to calculate the probability of drawing each ace, and then multiply all those probabilities together.

You have the probability for the first ace. Now what is the probability for the second ace? You have to remember that each time you pull a card, you have one less card in the whole deck, so there is one less possible outcome. Plus, you have already drawn an ace, so you have fewer aces.

4. What are the chances you draw a second ace?

5. And what about the probability of drawing a third and fourth?

Now multiply those four probabilities together.

6. What do you get in fraction form? What about in percent form?

10
SUDOKU

One day after school, Mason sees his dad playing a word game in the newspaper. Except it's not a word game—it's a number game. Mason's dad explains he's playing Sudoku, and then he tells Mason the rules.

The puzzle is arranged in a 9-square by 9-square grid. It is further divided into nine smaller squares, which are 3 squares by 3 squares of nine even smaller squares each (see the picture on the following pages). The game has some numbers filled in randomly around the grid. The goal is to fill in the rest of the squares. You can only use the numerals 1 through 9. You can't repeat any numbers in the small box of nine squares. You also can't repeat numbers in a row or column.

Mason sees that players have to be good at adding, and they have to be good at patterns and logic. See the Sudoku puzzle and the questions that follow to get a good idea of how Sudoku works.

		8						
4			1	5		3		
	2	9	4		5	1	8	
	4				1	2		
		6		2				
	3	2				9		
6	9	3		5		8	7	
	5		4	8				1
				3				

Try filling in the numbers on the third row. The third row already has a lot of numbers filled in for you.

You'll have to use a lot of trial-and-error. You can try out one number and work out a section from there. You might find it doesn't work, and have to replace that number and start the section over.

For the first square in the third row, you could try any number that isn't 1, 2, 4, 5, 8, or 9. You can also rule out 6, because 6 is already in the first column.

Next you could try the fourth square in the seventh row. The number you put there can't be 3, 5, 6, 7, 8, or 9 because they are already in that row. It also can't be 4 because 4 is in the same box of nine squares, and also in the same column.

Try to fill in the entire Sudoku puzzle. The first one you do might be hard, but don't let it frustrate you. If you keep going with them, you'll start to understand the logic behind them a little better.

11
BOWLING

Mason's friend Preeya invites him to go bowling. She knows he loves games and figures he would like to go bowling. She's right—Mason loves bowling, and he is pretty competitive about it.

Whenever he bowls, Mason likes to keep track of his own score. He knows the computer does it, but he likes to be sure he gets all the points he earned. He keeps track of everyone's score, not just his own.

Preeya's friend Jerome is also there. Jerome has never bowled before, so Mason explains the rules to him:

Each player gets two chances to bowl during each frame.

There are 10 frames in a game.

If you knock down all the pins, it's a strike. Strikes are worth 10 points, plus however many points you get in the next two rolls. You can't figure out your score for the frame when you get a strike until your next turn.

If you knock down all the pins using both chances, it's a spare. A spare is worth 10 points plus however many points you get on your next roll.

If you're left with at least one pin at the end of the frame, your score for the frame is however many pins you knocked down.

You get an extra roll in the last frame if you knock all the pins down.

Figure out how to score the whole game on the following page.

▶		1	2	3	4	5	6	7	8	9	10	T	

Preeya goes first. She knocks down two pins on her first roll, and gets a spare on her second. She has to wait until she rolls again in her next turn to figure out her score. On her next roll, she rolls a strike!

1. What is Preeya's score for the first frame?

On Mason's first turn, he gets a strike. He can't write his score down yet, because he has to add the value of his next two rolls. On his next two rolls, he knocks first 1 pin down, then 7.

2. What is Mason's score for his first frame?

On Jerome's first turn, he knocks down 2 and then 5 pins.

3. What is his score?

Here is the rest of what happens in the game. Use this information to fill out the score card. Use an "X" for strikes and a "/" for spares.

Frame 1: Preeya-2, spare; Mason-strike; Jerome-2, 5
Frame 2: Preeya- strike; Mason-1, 7; Jerome- 7, spare
Frame 3: Preeya- 3, spare; Mason- 6, 3; Jerome- 4, 2
Frame 4: Preeya- 4, 0; Mason- strike; Jerome- 8, 0
Frame 5: Preeya- 7, 2; Mason- 2, spare; Jerome- 8, 1
Frame 6: Preeya-0, 4; Mason- 4, 3; Jerome- 2, 5

Frame 7: Preeya- 8, 1; Mason- 4, 5; Jerome- 1, 8
Frame 8: Preeya- 0, 3; Mason-2, 0; Jerome- 4, 4
Frame 9: Preeya- 9, spare; Mason- 0, 9; Jerome- 4, 0
Frame 10: Preeya- 7, 1; Mason-2, 7; Jerome-9, spare, 7

4. Who won the game? How many points did he or she have?

5. How many points did the other two players have?

12
POOL

Another game Mason likes to play is pool. He just started learning it from his sister Lora. She's pretty good at pool, and she knows lots of tricks.

She and Mason are playing pool in their basement one day. She decides to tell him some of her tricks. She tells him pool is all about geometry and angles. Angles are the spaces between two lines, measured in degrees.

If you hit a pool ball with the **cue ball** at a certain angle, it will bounce away in the same angle. For example, if you hit the ball straight on, at a 180 degree angle, it would move away from you at a 180 degree angle. That would be a good shot if you could line up the cue ball, the ball, and the pocket you're trying to hit it into.

When you're playing, you won't be able to measure out the angles exactly. But you will be able to **estimate** them. The next pages show you some angles, and how to use them to become a great pool player!

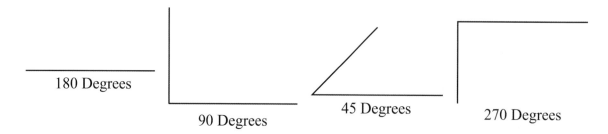

180 Degrees

90 Degrees

45 Degrees

270 Degrees

54

Mason starts out with an easy shot. There is a ball lined up with a pocket in the middle of the pool table's side, and the cue ball is in the same line.

Take a look at image A, where the white ball is the cue ball, and the black ball is what he is trying to put in the pocket. You can even draw the line from the cue ball, through the ball, and into the pocket to get a better sense of the angle.

1. What angle does his shot make with the ball if he lines his cue up with the cue ball, the ball, and the pocket?

You can also hit the cue ball so that it hits the ball and off the side at such an angle that it rolls into a pocket.

In image B, the dashed line shows the angle at which you should hit the ball on the right so it bounces off the side and into the pocket.

2. What angle is that closest to?

A third way to score is to bounce the cue ball off the side at such an angle that it bounces back and finally hits the ball, which rolls into the pocket.

The ball on the left of image B shows an example of this move.

3. What angle is it closest to?

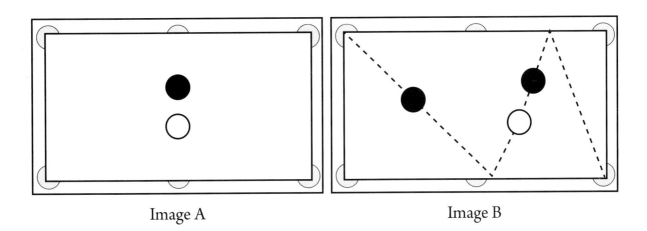

Image A   Image B

13
VIDEO GAMES: LINEAR ALGEBRA

Mason also loves to play video games. Whenever he's not playing a board game or a word game on paper (or on the computer), he's playing a video game.

One of his favorite video games is very simple. It involves flying a spacecraft around and blasting asteroids. If you don't blast the asteroids fast enough, they end up crushing the spaceship.

This video game uses something called linear algebra. And linear algebra involves things called vectors. Vectors are simply points with a direction. As the point moves, it turns into a line in the direction it's moving.

In Mason's video game, the moving spaceship is a vector. So is the asteroid. The next pages will help you understand vectors a little more.

The big dot represents the spaceship. If it moves in the direction of the arrow, what is its vector?

The spaceship moved 3 spaces to the left. In linear algebra, that means it moved -3 spaces (negative 3). Moving to the right is positive rather than negative. The spaceship also moved 2 spaces up. In linear algebra, that means it moved 2. Moving down instead of up is negative.

The way that vectors are written is: $(-3, 2)$

Horizontal movement is written first, then vertical movement.

1. What would the vector be if the spaceship moved 1 space to the right and 4 up? Draw it in the grid.

2. How about if the spaceship moved 2 spaces to the left and 1 down? Draw it in the grid.

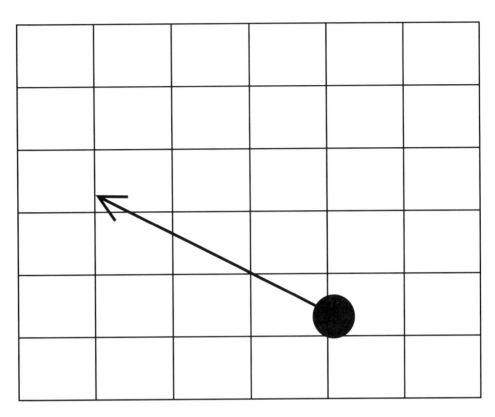

14

MORE VIDEO GAME MATH: SCIENTIFIC NOTATION

In the spaceship video game, Mason tends to get high scores because he's been playing for a long time. He has to work with some pretty big numbers. Every time he reaches 1,000 points, he gets a jewel. And every time he gets 10 jewels, he gets a planet.

The goal of the game is to collect an entire solar system's worth of planets, which, in the video game world, is one hundred planets. Mason wonders how many points winning equals. He can use math to figure it out—check out the following pages.

You want to find out how many points winning the whole game equals. First, find out how many points a planet is worth.

1 planet = 10 jewels, and 1 jewel = 1,000 points.

Multiply the number of jewels you need by the number of points in a jewel. So:

1. 10 x 1,000 =

Now multiply the number of points per planet by the number of planets it takes to win the game, which is 100.

2. How many points does it take to win?

You might notice there are an awful lot of zeroes in your answers. Math gives you a handy way to write big numbers with lots of zeroes, called scientific notation. It works like this:

$$10^1 = 10$$
$$10^2 = 100$$
$$10^3 = 1,000$$

Now fill in the rest of the pattern:

$$10^4 =$$
$$10^5 =$$
$$10^6 =$$
$$10^7 =$$
$$10^8 =$$

3. How would you write the number of jewels it takes to win the video game in scientific notation?

You can also easily multiply and divide numbers written in scientific notation. Just add the exponents (the little numbers) together for multiplication, and subtract for division.

The equation to find the total number of points to win then becomes:

$$10^1 \times 10^3 \times 10^2$$

Add the exponents together because you are multiplying.

4. Do you get the same answer as before?

15
PUTTING IT ALL TOGETHER

Mason never realized how connected games and math are. He likes both, so it makes sense to him that they have a lot to do with one another.

You can see how useful math is when playing games. Math helps you understand games better, and can even help you win! See how much you remember from all the games Mason has played.

1. What is the probability of rolling a 1 or a number greater than 4 on a six-sided die?

    Can you reduce that fraction? If so, what is its reduced form?

    What is that probability in decimal form?

    And in a percent?

2. In Monopoly, you are counting up all your money at the end of the game. Your opponent has $2,178. You have two $500 bills, six $100 bills, nine $50 bills, seven $20 bills, two $10 bills, and eleven $1 bills.

    Did you win?

    If so, by how much? If not, how much more did you need to win?

3. What is the area of a game board that is a rectangle that measures 14 inches by 22 inches?

4. You are playing multiplication War and you put down a queen and an eight. Your opponent puts down a jack and a nine.

   Who wins?

5. What is the probability that you draw a red ace out of a 52-card deck?

   What is the reduced fraction probability?

6. You are bowling and get a spare in one frame, followed by 8 pins and 1 pin in the next frame. Your friend gets a strike in one frame, followed by 2 pins and 4 pins.

   Who gets the most points for the first frame?

7. What is the vector of a video game character that runs 4 spaces left on a grid and jumps 7 spaces up?

ANSWERS

1.

1. 6
2. 1
3. ½; you can also say the probability is one-half.
4. ¹⁄₁₂ is smaller, because there are more outcomes, and less of a chance any one number will be rolled. ¹⁄₂₀ is even smaller, because there are even more outcomes.
5. ⁶⁄₂₀, or ³⁄₁₀

2.

1. 0.17
2. 17%
3. ⁴⁄₆, ²⁄₃

4. 0.67
5. 66.67%

3.

1. Row G, Column 5; no, he missed the ships.
2. B6, B8, A7, C7; they are all next to his correct guess.
3. ¼

4.

1. $1,500 x 5 = $7,500
2. Two $500 bills, four $100 bills, one $50 bill, one $20 bill, two $10 bills, one $5 bill, and five $1 bills.
3. $125
4. Yes—Vermont Avenue and Connecticut Avenue.

5.

1. A, G, P, W, K, X, Q
2. 44 points.

6.

1. 24 inches
2. 1 ½ inches
3. 3 inches; no
4. 1 square foot, 144 square inches.
5. 4 square feet, 576 square inches.
6. No, it is four times as big.

7.

1. Less: ⅟7, ²⁄7; More: ⁴⁄7
2. Rock = ²⁄7, Paper = ³⁄7, Scissors = ²⁄7
3. Less: ²⁄7, More: ³⁄7

Turn, Who Won

1, Michael
2, Tie
3, Mason
4, Mason
5, Mason
6, Tie
7, Michael

4. Mason won, 43% of the time.

1. Diego is winning. He has 6 cards from these rounds, and Mason has 4.
2. Mason
3. Diego
4. Diego has 4 more cards than Mason.

Mason	Diego
6 x 9 = 54	queen x 2 = 24
king x ace = 13	9 x 4 = 36
3 x 5 = 15	ace x 5 = 5
7x jack = 77	10 x 8 = 80
2 x 7 = 14	king x jack = 143

1. $^{26}/_{52}$, ½
2. $^{12}/_{52}$, $^{3}/_{13}$
3. $^{4}/_{52}$ (or $^{1}/_{13}$)
4. $^{3}/_{51}$
5. Third: $^{2}/_{50}$, Fourth: $^{1}/_{49}$
6. $^{24}/_{6,497,400}$, 0.000369378%

10.

3	1	5	8	2	7	9	4	6
4	6	8	9	1	5	7	3	2
7	2	9	3	4	6	5	1	8
9	4	6	5	3	8	1	2	7
5	7	1	6	9	2	4	8	3
8	3	2	1	7	4	6	9	5
6	9	3	1	5	1	8	7	4
2	5	7	4	8	9	3	6	1
1	8	4	7	6	3	2	5	9

11.

1. 10 + 10 = 20
2. 10 + 8 = 18
3. 7
4. Preeya won with 108 points.
5. Mason had 105 and Jerome had 89.

12.

1. 180 degrees.
2. 45 degrees
3. 90 degrees.

13.

1. (1, 4)
2. (−2, −1)

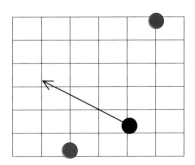

14.

1. 10,000
2. 1,000,000

   $10^4 = 10,000$
   $10^5 = 100,000$
   $10^6 = 1,000,000$
   $10^7 = 10,000,000$
   $10^8 = 100,000,000$

3. 10^6
4. yes

15.

1. ³⁄₆; Yes, ½; .50; 50%
2. Yes; You won by $43.
3. 14 x 22 = 308 square inches
4. Your opponent wins.
5. ²⁄₅₂; ¹⁄₂₆
6. You get the most points: 18.
7. (−4, 7)